PHYLOGEOGRAPHY: IT'S IMPORTANCE IN INSECT PEST CONTROL

INSECTS AND OTHER TERRESTRIAL ARTHROPODS: BIOLOGY, CHEMISTRY AND BEHAVIOR

Additional books in this series can be found on Nova's website under the Series tab.

Additional E-books in this series can be found on Nova's website under the E-book tab.

INSECTS AND OTHER TERRESTRIAL ARTHROPODS:
BIOLOGY, CHEMISTRY AND BEHAVIOR

PHYLOGEOGRAPHY: IT'S IMPORTANCE IN INSECT PEST CONTROL

M.D. OCHANDO

A. REYES

D. SEGURA

AND

C. CALLEJAS

Nova Science Publishers, Inc.
New York

For permission to use material from this book please contact us:
Telephone 631-231-7269; Fax 631-231-8175
Web Site: http://www.novapublishers.com

NOTICE TO THE READER

The Publisher has taken reasonable care in the preparation of this book, but makes no expressed or implied warranty of any kind and assumes no responsibility for any errors or omissions. No liability is assumed for incidental or consequential damages in connection with or arising out of information contained in this book. The Publisher shall not be liable for any special, consequential, or exemplary damages resulting, in whole or in part, from the readers' use of, or reliance upon, this material. Any parts of this book based on government reports are so indicated and copyright is claimed for those parts to the extent applicable to compilations of such works.

Independent verification should be sought for any data, advice or recommendations contained in this book. In addition, no responsibility is assumed by the publisher for any injury and/or damage to persons or property arising from any methods, products, instructions, ideas or otherwise contained in this publication.

This publication is designed to provide accurate and authoritative information with regard to the subject matter covered herein. It is sold with the clear understanding that the Publisher is not engaged in rendering legal or any other professional services. If legal or any other expert assistance is required, the services of a competent person should be sought. FROM A DECLARATION OF PARTICIPANTS JOINTLY ADOPTED BY A COMMITTEE OF THE AMERICAN BAR ASSOCIATION AND A COMMITTEE OF PUBLISHERS.

Additional color graphics may be available in the e-book version of this book.

LIBRARY OF CONGRESS CATALOGING-IN-PUBLICATION DATA

Phylogeography : it's importance in insect pest control / authors, M.D. Ochando ... [et al].
    p. cm.
  Includes index.
  ISBN 978-1-61209-371-0 (softcover)
  1. Phylogeography. 2. Insect pests--Control. I. Ochando, M. D.
  QH84.P499 2011
  576.8'809--dc22
                        2011003551

Published by Nova Science Publishers, Inc. † *New York*

CONTENTS

PREFACE

Phylogeography involves knowledge of the spatial distribution of related individuals and historical information on the relationship within and among populations and species. The phylogeography of many groups has been studied over recent decades, and this field of knowledge is now becoming important in solving the problems of pest control in agriculture and forestry. An understanding of the nature of the genetic variation within and between pest populations is of paramount importance in the design of pest control programmes and their success – and successful programs are certainly needed since increasing levels of trade and passenger travel, the growth of new plant species in new regions and climate change are all assisting the spread of economically important insect pests.

Molecular genetics is now providing us with new and more sensitive tools for developing appropriate insect control and eradication strategies. Our group is studying the genetic variation of tephritids and whiteflies, both of which are important pests of agricultural and ornamental plants in the Mediterranean region (ecosystems). The information presented here is a summary of some of the results obtained in enzyme electrophoresis (MLEE), abundant soluble protein content, random amplification of polymorphic DNA (RAPD-PCR), intermicrosatellite (ISSR), restriction fragment length polymorphism (RFLP) and mtDNA sequencing analyses. These data document the phylogeographic structure of these taxa's populations, and provide information regarding their bioinvasion and colonisation of new areas, their geographic variation, and gene flow among regions. Such information could be helpful in the taking of pest management decisions, and highlights the need for the coordination of control programs between regional, national and international authorities.

Chapter 1

THE PROBLEM

Insect pests are a growing problem worldwide; increasing levels of trade and passenger travel, the growth of new plant species in new regions and climate change are all assisting the spread of economically important insect pests. These pests cause tremendous harvest and economic losses, including prevention, quarantine, and eradication costs. Globally, the cost of the damage caused by invasive species has been estimated at close to 5% of the world's GDP, and in developing countries, where agriculture accounts for a higher proportion of the GDP, the negative impact of invasive species can be even greater (CABI, http://www.cabi.org).

In the past, synthetic pesticides (insecticides, fungicides and herbicides) seemed to be the solution to pest control and were seen as an integral part of agricultural intensification. The "green revolution" understood pesticides to be essential in improving harvests, and government instruments were made available to help make them accessible to farmers. However, just a few decades later the indiscriminate use, misuse and overuse of pesticides has led to increased resistance on the part of the pests they were meant to kill. Pesticide residues in food have become an important human health concern, and the pollution of the environment by these agents has led to concerns over the health of ecosystems and the loss of biodiversity. People are now realizing that while pesticides can be successful in the short term, they can cause many more problems than they solve in the long term. Pesticide use is now more controlled, with national and international laws establishing strict controls that favour the reduced use of pesticides, and the European Union is strictly controlling the quality of imported fresh food, meaning exporting countries must apply pesticide use controls. Nonetheless, these agents are still employed in huge amounts around the world.

The indiscriminate use of pesticides has required changes be made in plant protection strategies. Indeed, since the mid 1960s the FAO has been advocating Integrated Pest Management (IPM) as the preferred pest control strategy. IPM, as defined by the Dictionary of Biological Control (Coombs and Coombs, 2003), is "a system for controlling pests that is based on the combined use of a range of different methods (e.g., biopesticides, biocontrol agents, mating disruption, trapping and crop rotation, etc.) in order to minimise the use of chemical pesticides". IPM methods try to be more environmentally sensitive in their approach to pest management than other classical methods. IPM programs have been numerous and have been implemented in many developed countries. However, they face more problems in developing countries since they require more than the involvement of farmers, such as field staff from national and local governments and non-governmental institutions. Together these have the job of enhancing ecological awareness, of making evident the need for long term economic benefits, and of promoting environmental and human health safety. In summary, IPMs promote the growth of healthy crops by encouraging natural pest control mechanisms (FAO, http://www.fao.org/) and in so doing try to cause the least possible disruption of agro-ecosystems.

Later on, a derived system, Area-Wide Integrated Pest Management (AW-IPM) seeks to ensure protection from pests over very large regions, and so avoid new invasions. Conventional insect control has been, in general, short-term and small-area in its thinking, with little planning, and has involved the use of simple technology. However, such programs have commonly run into trouble and have required a move to area-wide insect control involving large numbers of producers and crops. The aim of area-wide control is to reduce the pest population within a large target area to a non-economically important level. This is accomplished by attacking the entire insect pest population in the target area (Tan, ed., 2000; Vreysen *et al.*, eds., 2007). To increase efficacy, high technology systems that can reduce costs and environmental problems are required.

An important point to address from the beginning in AW-IPM is the selection of the target pest. Many species might be candidates for control, but selection must start with those species that cause the greatest damage and that have the widest distribution, e.g., tephritids such as *Ceratitis capitata* and *Bactrocera oleae*, etc. As has been pointed out, IPM is a defensive procedure practised from a field point of view basis, but now, for AW-IPM to have the best chances of success, information on the structure and dynamics of populations and on the phylogeography of those populations is necessary (Cox, 2007).

Back in 1964, DeBach defined Biological Control (BC) as "…the action of parasites, predators or pathogens in maintaining another organism's population density at a lower average than would occur in their absence". Nowadays, BC is considered part of IPM and is becoming more popular, in part because of the surge in ecological agriculture and the increasing demand for its products. These days the definition has changed somewhat to reflect a more applied concept and has become "the use of predatory and parasitic insect species (termed natural enemies or beneficials), or natural products consisting of or derived from microorganisms, against pests and disease of crops" (Coombs and Coombs, 2003). For a good review of Biological Control see Bellows and Fisher (1999). However it should not be forgotten that, in many cases, chemical pesticides may also be required for the pest density to be brought a level at which BC is feasible.

The ultimate aim of BC is to maintain pest populations at a low level or eradicate them, thus reducing the risks of survival of sources of later reinfestation. However, BC requires more complete and intensive management and planning, the training of personnel, and it can take more time to achieve results. Some major successes has been achieved with BC, but the outcome of its use is normally far from predictable. Less than 30% of all attempts to achieve the control of pests via the introduction of its natural enemies are successful (Greathead and Greathead, 1992; Unruh and Woolley, 1999).

Currently, IPM, AW-IPM and BC are all implemented by governments and international organizations (e.g., the FAO, IOBC, International Organization of Biological Control, CABI, the Centre for Agriculture and Biosciences International [formerly the Commonwealth Agricultural Bureaux International]), and different governmental and non-governmental authorities are working to make farmers and the public aware of its possibilities, organizing meetings to discuss its use, and developing research projects, etc. Such is the case of the International Organization of Biological Control (IOBC, which is divided into a number of working groups and regional sections) which "promotes the use of sustainable, environmentally safe, economically feasible and socially acceptable control methods of pests and diseases of agricultural and forestry crops", and encourages collaboration in the development and promotion of biological and integrated production systems. The IOBC trains people in and informs people about biological methods of control, as well as the use of chemicals, within an IPM context. It also produces guidelines for the integrated production of crops, develops and standardises methods of testing the effects of pesticides on beneficial species, and organises scientific meetings and research on insect pests (IOBC: http://www.iobc-wprs.org/). Phylogenetic studies are now more commonly reported in such meetings.

Chapter 2

THE CHALLENGE

The EU is a major player in global agricultural trade; indeed, it is the largest importer and the second largest exporter of foodstuffs, it plays a leading role in establishing global trade agreements via the World Trade Organisation (WTO) and the European Commission of Agricultural and Rural Development (via the Common Agricultural Policy [CAP]), and it promotes sustainable agriculture in a global environment. It should be remembered that 90% of the land within the EU is either farmland or forestry. Farmers still receive subsidies, but in compensation they have to respect environmental safety, food safety and phytosanitary and animal welfare standards. The CAP not only ensures standards for EU farmers, it improves the quality of Europe's food (helping to guarantee food safety) and ensures that the environment is protected for future generations (http://ec.europa.eu/agriculture).

The European Commission is working on a strategy to reduce the impact of pesticides on human health and the environment. In fact, the last assessment on the European environment (The fourth assessment, 2007) highlights that "the historic impact of agriculture on landscapes and biodiversity was positive, but modern, intensive agriculture is often a threat to biodiversity. Agriculture has a negative influence on the environment through its use and pollution of resources such as air, water and soil". Fortunately, pesticide use appears to have been decreasing throughout the region since 1990, and the trend is towards the complete elimination of these chemicals.

While authorities are legislating for a more rational and safer use of pesticides, trying to enhance the development and current acceptance of IPM and BC, the conviction exists that better scientific knowledge of insect pests and their natural enemies will lead to more effective biological control. Indeed, agricultural associations are now demanding the EU legislate on a scientific basis. In this regard, molecular methods may provide us new characters of study to diversify our knowledge of the phylogenetic relationships of pests, to identify different biotypes and better understand their ecology and population structures, etc., and in so doing contribute towards better biological control and better legislation. In other words, from research to implementation.

Chapter 3

THE TOOLS: BIOGEOGRAPHY
AND MOLECULAR GENETICS

Phylogeography involves knowledge of the spatial distribution of related individuals and historical information on the relationship within and among populations and species (Avise, 1994, 2000; Lomolino *et al.*, 2006). The phylogeography of many groups has been studied in recent decades, but only recently has this field of knowledge started to become important in the study of insect pests in agriculture, ornamental plants and forestry. Detailed knowledge of the biology, genetic structure and geographical variability of a given pest species is a prerequisite in planning strategies for its quarantine, control or eradication (Roderick and Navajas, 2003). For example, the reconstruction of the histories of populations can be important in identifying the natural enemies of pests that can be used in biological control. Further, the identification of pathways of anthropogenically mediated introductions can assist in international efforts to limit the spread of non-indigenous pests. Different insect pest studies have demonstrated the value of knowledge on population genetics in pest management (Villablanca *et al.*, 1998; Gasperi *et al.* 2002; Meixner *et al.*, 2002; Ochando *et al.*, 2003a, b, etc.).

The role of molecular genetics technologies in studies related to population structure and the dynamics of insect pest species has increased rapidly during the last two decades. Rapid improvements in resolution, reproducibility, time, costs and standardization, plus recognition from insect control managers and authorities regarding the importance of such studies, is leading to the general acceptance that phylogenetic knowledge is a cornerstone in our fight against insect pests.

DNA methodologies in general, and those based on the polymerase chain reaction (PCR) in particular, are currently being used to answer a range of questions and are contributing significantly to our knowledge of population structures and dynamics, genetic mapping and phylogeny etc. (Loxdale and Lushai, 1998; Avise, 2003, 2004; Severson *et al.*, 2001; Heckel, 2003; Mendelson and Shaw 2005; Behura, 2006). They are also supplying more applied knowledge on insect/plant/pathogen interactions, insecticide resistance, mating behaviour related to sterile insect technique (SIT), and on the predators and parasites of pest species, etc. The use of both kind, nuclear and mitochondrial, markers have contributed to advances in our knowledge, and studies on amplified fragment length polymorphism (AFLP), random amplified polymorphic DNA (RAPD) and microsatellites, and the use of many others types of PCR technology (S-SAP, SNP, EPIC etc.), are slowly becoming popular tools for tackling insect pest problems (for a review of the use of molecular markers in insect studies see Behura, 2006). Advances in the inference of phylogenetic relationships from molecular data, however, require the use of appropriate statistical tools, and advances are also significant in this way (Swofford *et al.*, 1996; Nei and Kumar, 2000; Felsenstein, 2004).

Analysis of molecular data in a phylogenetic context requires that one be aware of the particular properties of the different types of data and of the different analytical approaches to data analysis. As have been marked by Unruh and Woolley (1999), three questions must be asked when one is contemplating a phylogenetic study involving molecular markers. The first one is about the rate of evolution of the marker, is it a rate appropriate for the biological system under study?. One of the reasons that mtDNA has been used so successfully in many studies of infraspecific phylogeography is that restriction endonucleases generate useful and informative variation at the level of local populations. Second, are we making the comparisons between homologous features in different taxa? do we know sufficiently the source of variation of the used marker?. And third, are our methods of analysis sensitive enough to detect variation in rates of evolution among different marker systems, or between different taxa when using a single marker system?.

Phylogenetic methods have several distinct advantages. Certainly they provide a simple and rigorous context in which to directly compare information from various sources e.g., different sequences of DNA and types (from a selective point of view) of DNA fragments, allozymes and morphology. Another advantage is that such information can provide direct evidence supporting the existence of particular clades (groups or lineages).

In sum, the biogeographic information obtained through the use of these molecular technologies is probably now one of our most powerful tools in the eventual control of insect pests.

Chapter 4

SPECIFIC QUESTIONS

The agricultural, forestry and ornamental plants importance of insect pests highlights the need to understand pest population structures and dynamics. Phylogeographic genetics is helping to answer basic questions regarding bioinvasions, the geographic sources of invasions, population genetic structures, gene flow, host/geographic biotypes and colonisation routes etc. This book reports some of the results (partly already published) obtained by our group in these areas regarding pests of Spain and the Mediterranean Basin in general. The overall aim of our research was to use DNA marker methodologies to study the phylogeography, geographic population structure, gene flow, molecular systematic and phylogenetics of insect pests, and to make known the importance of this kind of work in the outcome of pest control management actions.

Chapter 5

BIOINVASIONS: WHITEFLIES
IN THE CANARY ISLANDS

The invasion of an area by an exotic species could be a unique event or the outcome of waves of re-introduction?. Can the origin of the invader be determined?

The increasing level of trade, the cultivation of plant species in new regions and climate change may all be significant aspects related to the invasion a territory by exotic pests. Spain is the gateway between Europe and Africa, which lies just 14 km away across the Straits of Gibraltar. Spain's biodiversity is the highest in Europe – but it is also a part of the continent likely to be most affected by climate change. Further, it is the most important point of trade and travel between Europe and Central and South America. Together, these factors leave Spain particularly vulnerable to bioinvasions. Some recent examples include the arrival of new and more pathogenic biotypes of the whitefly *Bemisia tabaci* in the country's greenhouses, and the appearance –now or in a short time expected- of *Bactrocera zonata*. In addition, the European Food Safety Authority (EFSA) has confirmed that the pest control measures applied by the EU to citrus fruits imported from South Africa are insufficient, with the consequent risk of the introduction of the fungus *Guignardia citricarpa* (the causal agent of black spot). Moreover, in the Mediterranean region of Spain, three other exotic pests have been detected over the last two years. Tougher controls are therefore required - controls that need to be based on better scientific knowledge. Phylogeographic information could be of paramount importance in this respect.

The whitefly *Aleurodicus dispersus* (Russell 1965, Hemiptera, Aleyrodidae) is a highly polyphagous, haplodiploid species that has been recorded on about 100 species of plants belonging to nearly 30 families, including many ornamental, vegetable and fruit crops (Waterhouse and Norris, 1989). This species, native to the Caribbean region, was first recorded in the Canary Islands in 1965 on *Schinus terebinthifolius*. Since then, it has spread throughout the archipelago (Hernández-Suarez *et al.*, 1997; Beitia, 1998; Martin *et al.*, 2000), and since the 1990s it has been an important pest on ornamental and tropical crops. The nature of the invasion process and the geographic source of the original invading population were questions to which control managers needed answers – answers that the phylogeographic studies undertaken by our group helped provide (Callejas *et al.*, 2005).

To investigate recent bioinvasions of exotic species, markers revealing the variation among recently diverged populations are needed. We decided to use RAPD markers. The Random Amplified Polymorphic DNA technique (Williams *et al.* 1990; Welch and McClelland, 1990) involves the amplification of random polymorphic segments of genomic DNA using single primers of arbitrary nucleotide sequence. Like any other genetic marker, RAPD markers have some limitations, including limited reproducibility and marker dominance. Among the DNA fingerprinting techniques, however, RAPD requires the least economic input, the smallest amount of laboratory equipment and labour, its usefulness in insect population studies is well known, and most importantly, RAPD markers show a very high degree of variability among samples at the infraspecific level. RAPD-PCR was therefore used to study seven samples of *A. dispersus* from different hosts on different islands of the Canary group (Figure 1). Fifty adults from each population were analysed, along with 14 pupae from Costa Rica, which were used as an outgroup.

Genomic DNA was extracted according to our own protocol based on that of Higuchi (1989) with some modifications. Six primers - A03, A13, B08, C08, F06 and F08 - (Operon technologies, Alameda, USA) were used. The amplification reactions were performed according to Williams *et al.* (1990) with minor modifications. The thermocycler program used was: preheating at 94°C for 5 min, 45 cycles of amplification (1 min at 94°C, 1 min at 36°C, 6 min at 72°C) and a final extension step of 6 min at 72°C. Along with a standard molecular weight marker (100 bp Ladder Plus, MBI Fermentas) the PCR amplification products were loaded onto 2.0% agarose gels in a buffer solution (1X TAE) containing ethidium bromide (0.5 µg/ml). All reactions were performed following a strict protocol with standardised conditions, repeating each amplification reaction at least twice. All the amplification products obtained were reproducible and consistent. Figure 2 shows an example of the RAPD profiles obtained with primer A03.

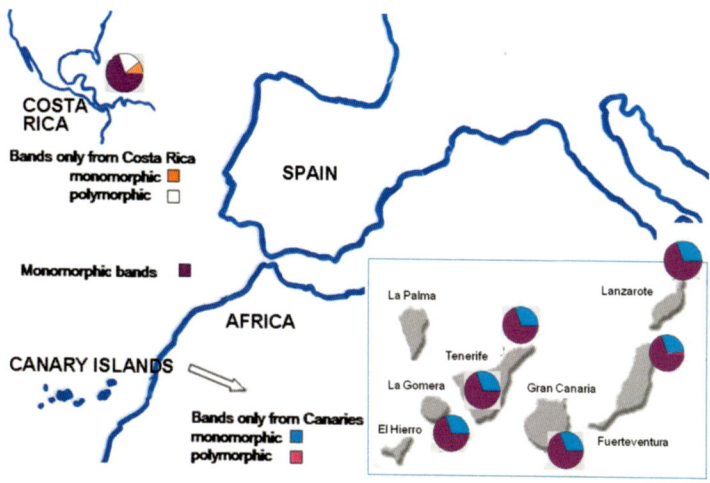

Figure 1. Collection sites and frequencies and distribution of monomorphic and polymorphic RAPD bands for populations of *A. dispersus* from the Canaries and Costa Rica.

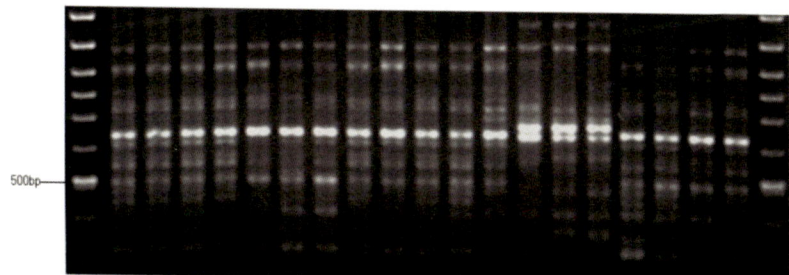

Figure 2. RAPD profiles of the *A. dispersus* flies analyzed with primer A03. First and last lanes contain a 100bp ladder molecular weight marker and the remaining correspond to specimens from different populations.

A total of 68 bands were scored, their size ranging from 240 to 1400 bp. No differences in RAPD patterns were found among the samples from the different Canary islands, except for one exclusive band present in just one fly from Lanzarote (Figure 1).

Unexpectedly, the number of generations elapsed from the first detection of the pest in the Canary Islands (9-12 per year, Hernández-Suarez, 1999) seems not to have been enough to generate variability or to provide evidence of possible selection effects. The founder effect therefore appears to have been very strong. Haplodiploid species may commonly show reduced genetic variability compared to those in which both sexes are diploid; in the haploid sex all loci would be subject to selection. Our results show there to be no genetic differences in *A. dispersus* inhabiting the different islands of the Canary archipelago. Neither were any differences observed relating to geographic origin, host/plant relationship or year of sampling. This very low level of genetic variation, with all populations showing the same DNA bands, supports the hypothesis of a single colonisation event by a small number of *A. dispersus* whiteflies followed by recent dispersion from the introduction point.

The results for the Costa Rican outgroup flies were quite different from those of the Canary Island populations. Thirty six of the total 68 bands scored (53%) showed fixed differences between the samples from the Canary Islands and Costa Rica. 'Regional' diagnostic bands, i.e, bands of DNA present in all whiteflies from the Canary Islands but absent in those from the outgroup and *vice versa* were also found: 11 in the Costa Rican whiteflies (including five that were polymorphic) and 15 in those from the Canary Islands. Thus, the flies of these two regions are quite well differentiated. Costa Rica is therefore not the source of the Canary Islands infestation.

In summary, the results of these RAPD phylogeographic studies suggest that the colonisation of the Canary Islands by *A. dispersus* began with a single colonisation event followed by dispersion, and that these invaders did not come from Costa Rica.

Chapter 6

COLONISATION PATTERNS:
THE IBERIAN PENINSULA IN THE
EXPANSION HISTORY OF THE MEDFLY

Can the colonisation process of a pest that has been established in a region for a long time be known? Can origin and expansion routes of such a pest be determined?

Intraspecific historical phylogeography - the study of the geographic distribution of a genealogical lineage (Avise *et al.*, 1987; Avise, 1991a, b, 2000) - was introduced at the end of 1970s, mostly involving mtDNA studies (especially restriction fragment length polymorphism [RFLP] studies). This technique was much more powerful and informative than those previously used, such as allozyme studies or even the measurement of nuclear sequence variation. The maternal inheritance of mtDNA, its rate of evolution, its lack of recombination and introns, its extensive polymorphism and the relative ease of its purification offer clear advantages in intraspecific phylogeography´ studies (Avise, 1994; Swofford *et al.*, 1996; Unruh and Woolley, 1999). Indeed, mtDNA markers have been among the most useful in population and phylogeographic analyses of all species (Avise, 1986; Hillis *et al.*, 1996; Zhang and Hewitt, 1997; Behura, 2006).

Along with MLEE, abundant soluble protein, RAPD-PCR and ISSR techniques, we used RFLP to study the colonisation history of *C. capitata* in the Iberian Peninsula. From a phylogeographic point of view, and for determining the colonisation route followed by the pest, the results shown below are some of the most important obtained.

The family Tephritidae (Tephritidae, Diptera), with more than 4000 fly species, is probably one of the most economically important of all the

Diptera. The majority of its species are fruit pests, and the Mediterranean fruit fly, *Ceratitis capitata* (Wiedemann) (Diptera: Tephritidae) is among the world's most destructive and economically devastating pest species (Figure 3). It is responsible for direct economic losses in fruit production, and is the focus of considerable and costly detection and eradication programs in all countries where it is found (Dowel and Krass, 1992; Bohonak *et al.*, 2001; Silva *et al.*, 2003; Meixner *et al.*, 2002; Ochando *et al.*, 2007; Malacrida *et al.*, 2007; Aluja and Mangan, 2008). Over the last 150/200 years, the medfly has expanded rapidly via its own power of dispersal and via human-mediated transport from its putative source area in Central Africa (Hagen *et al.*, 1981) to almost all regions with temperate or tropical climates (the Mediterranean region, South Africa, Central and South America, and even Australia) (Fletcher, 1989). It has even recently been detected in North America. The original host of the species was *Argaria spinosa* (L.), but the medfly now infests more than 250 species and varieties of agriculturally important plants (Fimiani, 1989). It is suspected that the Iberian Peninsula has played an important role in the spread of *C. capitata* to the Mediterranean region (Hagen *et al.,* 1981) and possibly to some regions of America. Genetic studies of wild Spanish populations of *C. capitata* may help to reveal its pattern of invasion and, subsequently, assist in designing strategies for its control and eradication.

Over the last 10 years a number of authors have attempted to resolve the fine geographical scale of medfly invasions, and to determine the genetic structure of its populations using different kinds of genetic markers (MLEE: Reyes and Ochando, 1994; Baruffi *et al.*, 1995; Malacrida *et al.*, 1998; Ochando *et al.*, 2003a; microsatellites: Bonizzoni *et al.*, 2000, 2001, 2004; Meixner *et al.*, 2002; RAPD: Haymer and McInnis, 1994; Baruffi *et al.*, 1995; Reyes and Ochando, 1998; Gasperi *et al.*, 2002; intron loci: Gomulski *et al.*, 1998; Villablanca *et al.*, 1998; Davies *et al.*, 1999; He and Haymer, 1999; RFLP of mtDNA: Silva *et al.,* 2003; Reyes and Ochando, 2004, etc.).

However, as Davies *et al.* (1999) have stated *"... genetic analysis...should begin with mtDNA... "* We used RFLP of mtDNA to study the variation of Spanish populations of *C. capitata* (Reyes and Ochando, 1998b, 2004). The aim of the work was to provide information about the phylogeographic relationships of medfly populations in this region, and the colonisation route taken across the northern Mediterranean.

Flies from different parts of Spain were collected by harvesting infested fruit and allowing the larvae to pupate in the laboratory. Twenty isofemale lines survived under laboratory conditions during the experiment: five from a population from central Spain (40°27'N 3°49'W, about 600 km north of the

Straits of Gibraltar; known as population CENTRE), five from a population from eastern Spain (39°26'N 1°10'W, about 70 km from the Mediterranean coast and 700 km northeast of the Straits of Gibraltar; known as population EAST), and 10 derived from a population from southern Spain (36°30'N 5°20'W, about 70 km north of the Straits of Gibraltar; known as population SOUTH). Mitochondrial DNA was obtained according to Afonso *et al.* (1988) with minor modifications. RFLP was studied using 22 restriction endonucleases, 17 of which (*Asp* 718, *Bam*H I, *Bcl* I, *Bst*E II, *Cla* I, *Dra* I, *Eco*R I, *Eco*R V, *Hin*d III, *Hpa* I, *Pst* I, *Pvu* II, *Sac* I, *Sal* I, *Sma* I, *Xba* I and *Xho* I) recognized 6 bp sequences, and five of which (*Cfo* I, *Hae* III, *Hin*f I, *Hpa* II and *Rsa* I) recognized 4 bp sequences. The restriction fragments obtained were resolved in 0.8-1.5% agarose gels in the presence of TAE buffer containing ethidium bromide (0.5µg/µl). Lambda phage digested with *Hin*d III or *Bst*E II was used as a molecular marker.

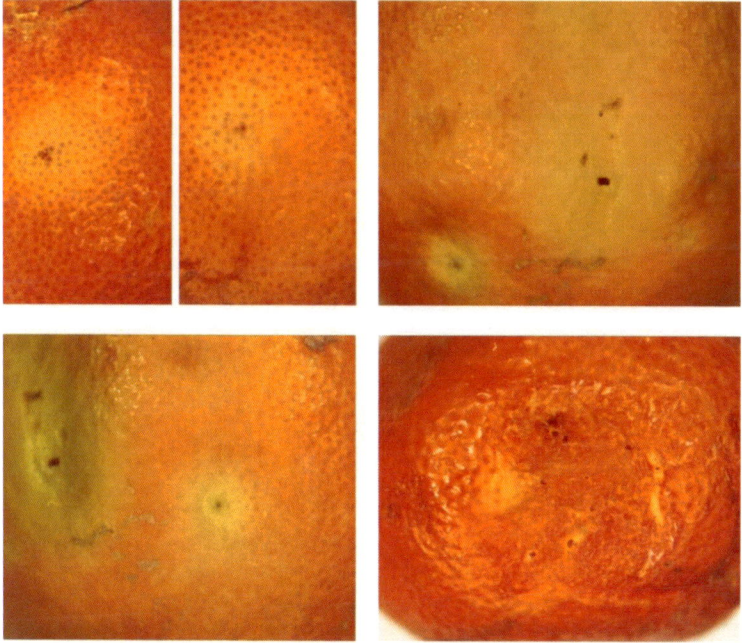

Figure 3. Infested citrus fruits by *Ceratitis capitata*.

Ten of these endonucleases (*Bcl* I, *Eco*R I, *Eco*R V, *Hin*d III, *Hpa* II, *Pvu* II, *Rsa* I, *Sac* I, *Xba* I and *Xho* I) yielded the same pattern for all the isofemale lines, five did not cut the *C. capitata* mtDNA (*Cla* I, *Hpa* I, *Pst* I, *Sal* I and *Sma* I), and the remaining seven enzymes (*Asp* 718, *Bam*H I, *Bst*E II, *Cfo* I, *Dra* I, *Hae* III and *Hin*f I) revealed RFLP. A total of 65 mtDNA restriction sites seen in the 20 isofemale strains of *C. capitata*; 59 (90.8%) were the same for all strains while the other six (9.2%) were polymorphic (Figure 4).

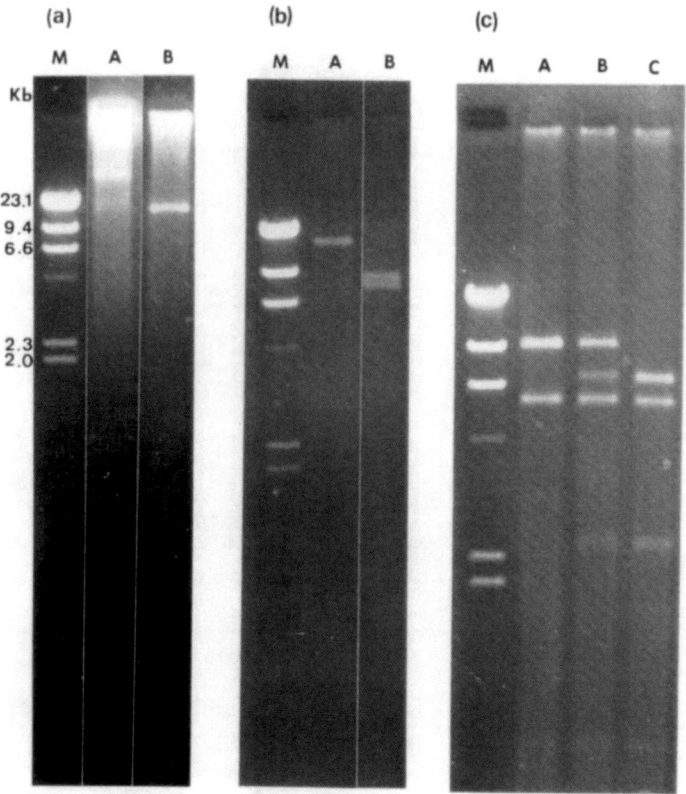

Figure 4. Restriction fragment length polymorphism profiles with restriction endonucleases *Asp* 718 (a), *Bst*E II (b) and *Hae* III (c) of *Ceratitis capitata* specimens. Molecular weight marker (M) corresponds to lambda phage cut with *Hind* III.

The combination of single patterns resulted in nine composite patterns or haplotypes, numbered I to IX. Based on a maximum parsimony network (see Reyes and Ochando, 2004, for a detailed explanation of the unrooted tree), haplotypes II and VII were found to be in central positions. Haplotypes

VII and VIII were the only two present in all three Spanish populations, while haplotypes I to VI were present only in population SOUTH, and haplotype IX was present only in population CENTRE. Thus, haplotype VII can be deemed the ancestral haplotype in Spain.

As mentioned above, haplotypes I to VI and IX were found only in single populations - SOUTH and CENTRE respectively. This suggests that they appeared after the colonisation of these areas. Moreover, the haplotypes present in more than one population (VII and VIII) appear with different frequencies (Figure 5). All the haplotypes (except one) were present only in population SOUTH. Maximum likelihood analysis (Nei and Li, 1979; Nei and Tajima, 1981) showed the nucleotide diversity to be highest in population SOUTH (0.30%), followed by CENTRE (0.24%), and finally by EAST (0.06%).

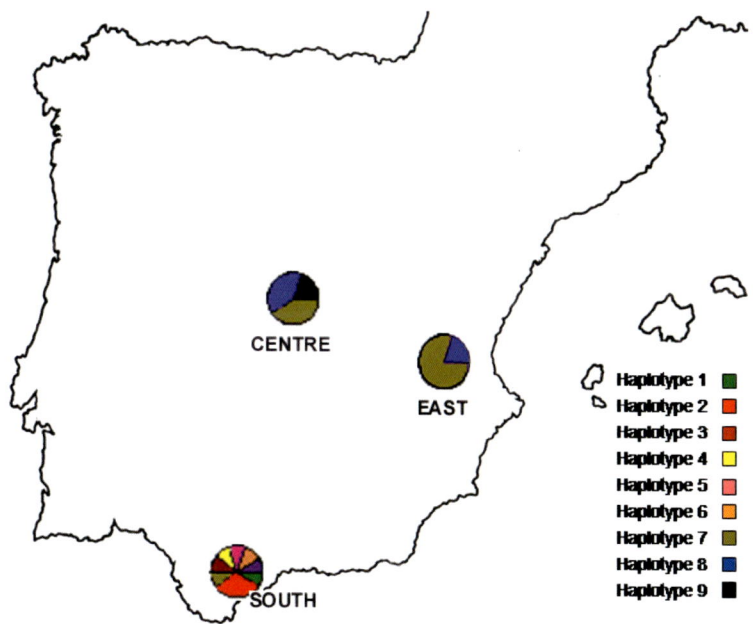

Figure 5. Population sampling sites and mtDNA haplotypes frequencies and distribution in the three populations of *Ceratitis capitata* analysed.

The colonisation of Spain by this pest is thought to have taken place via the Straits of Gibraltar, followed by its dispersion to more northerly regions (Hagen *et al.*, 1981). Accordingly, a decreasing level of variability from the source area towards the areas more recently colonised should be detected.

Indeed, the highest value of nucleotide diversity was found in population SOUTH (0.30%), which is geographically very close to the Straits of Gibraltar (about 70 km). Populations CENTRE and EAST had lower nucleotide diversity values (0.24% and 0.06% respectively). These results are in agreement with those obtained for the same populations when using molecular markers such as isozymes, when using RAPD-PCR (Reyes, 1995), or when examining the content of soluble proteins (Reyes and Ochando, 1998), and with those obtained in other Spanish populations of this species (Reyes and Ochando, 1994).

Medfly populations can be divided into three main categories according to their colonisation pattern: ancestral, ancient and new populations (Malacrida et al., 1992), corresponding to populations from Africa, the Mediterranean Basin and America, respectively. The question arises as to whether results from analyses involving the mitochondrial restriction sites in flies from these different regions are in agreement with the proposed colonisation process of this pest. In American populations, only one haplotype per geographical area has been found. In the case of Spanish populations, up to nine different haplotypes have been described, irrespective of sample size. The low degree of genetic variability in American populations (Sheppard et al., 1992; McPheron et al., 1994; Gasparich et al., 1995, 1997; Steck et al., 1996; Meixner et al., 2002) suggests the recent colonisation of America by African populations - both the African and American populations share a haplotype. African haplotypes are not seen, however, in Mediterranean populations. Although most Spanish haplotypes differ from the African haplotypes (Sheppard et al., 1992; Gasparich et al., 1997) by only one mutational step, indicating a possibly direct African origin, the large number of Spanish haplotypes would appear to indicate that the colonisation process is not as recent as in America. Haplotypes from Greece are not directly related to African ones but to Spanish haplotypes, suggesting that they might be derived from Spanish populations. However, this should be interpreted with caution since only one laboratory population from Greece has been analysed (Kourti, 1997). Overall, these findings are substantially in agreement with those obtained with other genetics markers and with the historical reports of movements of this pest (see below).

Isozyme and microsatellites analyses performed on African, American and Mediterranean populations (Huettel et al., 1980; Loukas, 1989; Gasperi et al., 1991; Malacrida et al., 1992, 1998, 2007; Reyes and Ochando, 1994, 1998; Baruffi et al., 1995; Bonizzoni et al., 2000, 2001, 2004; Meixner et al., 2002; Ochando et al., 2003a) have also revealed the greatest genetic variability (assessed by the proportion of polymorphic loci, the mean number of alleles per locus, and the average heterozygosity) to be shown by the populations of Africa, followed by those of the Mediterranean and finally those of America. Among the Mediterranean populations, those of Spain seem to show higher variability than those of Italy, followed by the Greek populations. The results of such macrogeographic analyses appear to be in agreement with one another whatever the molecular marker used. However, from a microgeographic point of view the data are not always so clear. For example, RAPD studies on Spanish populations of *C. capitata* show no clear quantitative nor qualitative distribution trends (Reyes, 1995). The explanation may lie in the high genetic flow among regions and the fact that consecutive generations of the fly need to feed on the fruit available at the moment. "Generalist" alleles are therefore required rather than "specialist" alleles that would adapt them to a specific kind of fruit. The non-existence of host biotypes supports such an interpretation (Ochando *et al., 2003b*)

In summary, the results obtained in mtDNA, isozyme, microsatellite, intron loci and RAPD analyses suggest that *C. capitata* has moved from its source area in Africa to Spain, across the Straits of Gibraltar, probably followed by eastward movements into other northern Mediterranean countries. With respect to the American populations, the low degree of variation (both in mtDNA and isozymes) and the results of intron loci studies (Villablanca *et al.*, 1998; He and Haymer, 1999; Davies *et al.*, 1999) seem to indicate an affinity of these populations with those of Africa - perhaps the result of a recent colonisation event directly from the source area. This hypothesis is in agreement with the dates on which the pest was first noticed in different countries. It was first reported as a pest in Spain in 1842 (De Breme, 1842 cited in Fimiani, 1989), in Italy in 1863 (Martelli, 1910 cited in Fimiani, 1989) and in Greece in 1915 (Papageorgiou, 1915 cited in Fimiani, 1989). It was then noticed in Argentina in 1905 (Gallo *et al.*, 1970) and in Hawaii in 1910 (Compere, 1912 cited in Headrick and Goeden, 1996). In 1975 it arrived in California (Carey, 1991), and in or before 1955 the medfly successfully established itself in Central America (Sheppard *et al.*, 1992). The progression of these dates is highly concordant with the data obtained with molecular markers.

Notwithstanding, there are other processes besides colonisation that could also influence the degree and/or kind of differentiation seen in the present populations: gene flow and selection. The existence of significant levels of gene flow between populations might limit the geographical differentiation seen between them, even when there is little or no selection. Trade, especially from the south and east to the centre of Spain is extensive, and human mediated movement of the pest between these regions should not be neglected.

In conclusion, the analysis of mtDNA variation in Spanish populations of *C. capitata* revealed them to show low to moderate nucleotide diversity, probably due to the "relatively" recent colonisation (less than two centuries) of Spain by this pest. It would seem, however, that Spanish populations have played an important role in the colonisation of the northern Mediterranean region. The colonisation of America by this species seems to have taken place later than the invasion of Spain and by a different route, probably directly from Africa. More information needs to be collected on *C. capitata* in Spain, not only through "traditional" RFLP analysis, but through a combination with other methods such as intron and ISSR analysis (this is now underway at our laboratory). This could provide more information on levels and patterns of polymorphism, which would be useful in the study of re-invasions and very recently colonised areas.

Chapter 7

POPULATION STRUCTURE: THE OLIVE FLY IN THE MEDITERRANEAN BASIN

Certain questions arise when considering insect pests with a wide range of distribution that have been established for a long time. Have they differentiated geographically? Is gene flow significant in the prevention of differentiation? Are there different selective pressures at work in different geographic areas?

The olive fruit fly, *Bactrocera oleae* (Gmelin), is a major olive crop pest. Its larvae are monophagous, feeding exclusively on olive fruits (see Figure 6). Crop losses include reduced harvests, reduced oil contents and reduced quality. In the Mediterranean basin, where 98% of the world's cultivated olive trees are found, production losses can reach more than 30%. Olive tree cultivation accounts for some 3.37% of the total agricultural production of the European Union (EU). Spain is the foremost olive oil-producing country in the world, followed by Italy and Greece (data from the FAO, 2004; http://faostat.fao.org/faostat).

The olive fruit fly is widespread throughout the Mediterranean and Middle East, and records of infestations go back some 2300 years (Ruiz, 1948). The pest is also found along the east coast and in the south of Africa, in India and Pakistan, and it was first detected in California in 1998 (Rice *et al.*, 2003). Olive fly researchers generally agree that this insect can survive and develop in any area of the world where olive trees - wild or cultivated - grow (Rice, 2000). Its control relies mainly on chemical treatments, sometimes applied over vast areas by aircraft, with the subsequent ecological and toxicological side effects such practices entail (Alberola *et al.*, 1999).

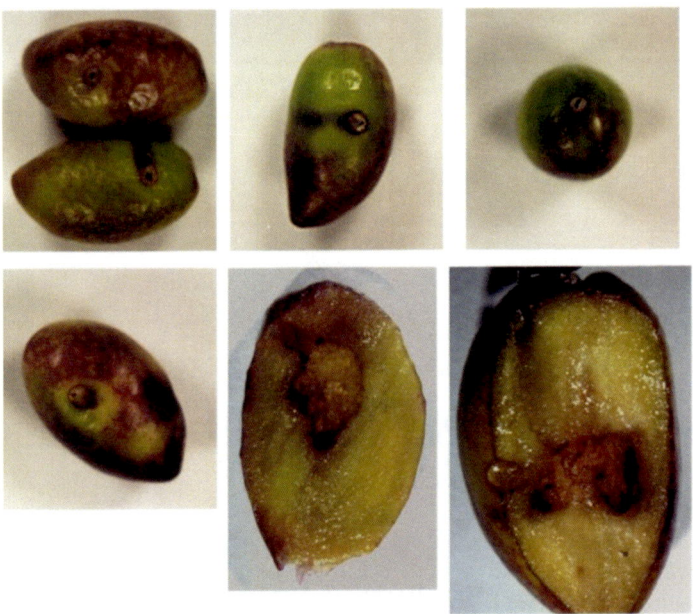

Figure 6. Infested olive fruits by *Bactrocera oleae*.

Given the economic importance of this pest, our knowledge of the olive fly is relatively extensive (see Bush and Kitto, 1979; Zouros and Loukas, 1989; Robinson and Hooper, eds., 1989; Ochando *et al.,* 1994; Barranco *et al.*, eds., 2004), but phylogeographic information is lacking; indeed, only a few recent reports exist (Callejas *et al.,* 1998; Ochando and Reyes, 2000; Ochando *et al.,* 2003a; Nardi *et al.*, 2005; Augustinus *et al.*, 2005). However, the efficient control of this pest requires information on its population structure, its patterns of colonisation, and the origin and spread of invading populations be known. Our group has published several papers on the genetic structure of the populations of the olive fly, mostly from the Mediterranean. Some of the results are summarized below.

In one study, 18 olive fly populations were examined, 13 covering the total range of the pest in the Iberian Peninsula (SP-1, SP-2,....SP-13 and POR), plus one each from Italy (ITA), Greece (GRE), Tunisia (TUN) and Israel (ISR) and California (USA). In all cases (except for the American population), 20 individuals per population were analysed through RAPD-PCR technique. The genomic DNA from individual flies was extracted according to Reyes *et al.* (1997), and DNA amplifications performed according to Williams *et al.* (1990) with minor modifications. Seven arbitrary sequence oligonucleotides (A-02, A-07, A-17, C-05, C-06, C-11 and C-18) from Operon Technologies (Alameda, CA, USA) were used in amplifications performed in an M.J. Research PT-100 thermocycler. The reaction conditions were as follows: preheating to 94° for 5 min, followed by 45 amplification cycles of 1 min at 94°C, 1 min at 36°C and 6 min at 72°C, and a final extension step at 72°C for 6 min. Each amplification reaction was performed at least twice: the results were consistently reproducible. The amplification products were separated according to their molecular size by electrophoresis in 2% agarose gels in the presence of TAE buffer (40 mM Tris-Acetate, 1mM EDTA pH 8.0) and ethidium bromide. A 100 bp ladder marker was used as a molecular size standard.

In general, the polymorphism detected for the olive fly is higher than that reported in RAPD studies of other insects (see De Sousa *et al.,* 1999; Lin *et al.,* 1999; Zitoudi *et al.,* 2001; Ochando *et al.,* 2003a). In our study a total of 115 bands were obtained, 98 of which were polymorphic and 17 monomorphic, and a high level of polymorphism was seen (mean 63%; range 51-70%) (Figure 7). The population diversity, as measured by Shannon's diversity index (H), ranged from 3.93 to 5.28. The literature (Zouros and Loukas, 1989; Ochando *et al.,* 2003a; Nardi *et al.,* 2005; Augustinus *et al.,* 2005) suggests high genetic variability to be characteristic of the species.

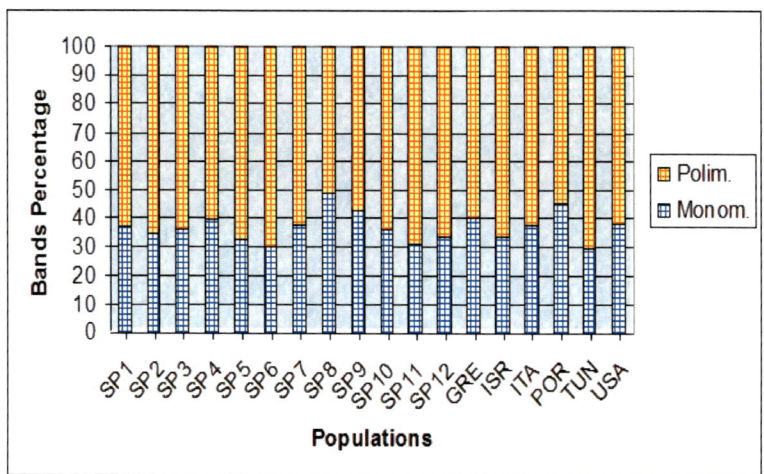

Figure 7. Percentage of monomorphic and polymorphic bands for each population of *B. oleae* sampled, indicated on the figure with different colours.

According to some authors a positive correlation is to be expected between the degree of environmental diversity of a species and its degree of genetic variability, and the potential for ecological heterogeneity to promote genetic diversity, and perhaps divergence, has been proposed (Abrahamson and Weis, 1997; Downie *et al.*, 2001). However, this does not seem to be the case for *B. oleae*. Indeed, comparisons with the information available for other Tephritidae underscores the idea that olive fly populations keep great genetic variation (Yong, 1992; Kourti, 2004; Callejas and Ochando, 2004). The high genetic variability of *B. oleae* populations is probably due to their large effective size. Olive groves cover wide expanses of territory; they can therefore maintain high population densities. Further, the species became established in the Mediterranean a long time ago. It is thought that it arrived in Europe more than two thousand years ago with the introduction of the olive tree from Western Asia and Africa by the Phoenicians (Ruiz, 1948).

The Californian population also showed high variability (polymorphism value 62% and the Shannon index: 3.93), which is of the same order as that seen in its European counterparts. These data indicate that this population is most likely the descendant of Mediterranean colonisers (as indicated by Nardi *et al.*, 2005) with a high degree of polymorphism, or that it has been established in America for some time.

However, all the primers used detected lower levels of genetic differentiation among the populations than within them. As much as 86% of the total diversity was attributable to diversity within populations, and just 14% to differences among populations. Similarly, AMOVA analyses showed that nearly all the total genetic variation to be maintained within populations (from 85.07 to 92.28% depending on the analysis). A random permutational test revealed that these variance components were all significant (p<0.001). For a detailed explanation of these data see Segura *et al.* (2007, 2008). These results agree with previous data recorded for this species (Zouros and Loukas, 1989; Ochando *et al.*, 1994; Callejas *et al.*, 1998), especially with the data on microsatellites and mitochondrial sequences reported by Nardi *et al.* (2005). Augustinus *et al.* (2005) reported the variation within populations to be 96.11% and the genetic distances between populations to be of the same order as in our work (except for their Cyprus samples) – although these authors draw different conclusions.

Gene flow must be responsible for the uniformity seen among the northern Mediterranean populations of the olive fly (Spanish populations plus Italian, Greek and even Israeli population). The existence of small and isolated populations would lead to divergence between populations and homogeneity within them. Conversely, the presence of large, interconnected populations would result in less interpopulational differentiation and greater diversity within these populations (Lin *et al.*, 1999). Different authors have reported adult olive fly movements ranging from 200 m in the presence of olive hosts, to as much as 4000 m if hosts need to be searched out. Indeed, dispersals of up to 10 km have been reported over open water in the Mediterranean (Rice, 2000). The passive transport of olive flies associated with human activities (such as transport and new plantations of olive trees) must be considered as well. An effective migration rate (Nm) of 1 is sufficient to prevent genetic differentiation among populations (Wright, 1931; Maruyama, 1970, 1972). The gene flow estimated for the Mediterranean populations in our work was more than 4, indicating that this may be an important factor influencing the genetic structure of *B. oleae* in this region where large populations have long been established.

The UPGM dendrogram (Figure 8) shows the relationships among the 18 populations of *B. oleae* analysed in this study. The bootstrap values were generally low (except for the first branching), supporting the idea that the majority of these populations are genetically so similar that they are difficult to separate. Notwithstanding, both the dendrogram and PCA analysis (Figure 8 and Figure 9) show the most southerly of the Mediterranean populations - the African population (TUN) – to differ significantly from the remaining populations. The American (USA) population at the next level of branching, form a group separated from the rest of the populations. Augustinus *et al.* (2005) reported the existence of subpopulations in the Mediterranean. However, in our case this conclusion is not statistically supported; the Italian, Greek, Israeli and Iberian populations cluster together. These results generally agree with those of Nardi *et al.* (2005) who used others markers (microsatellites and mtDNA sequences); these authors also found two genetics groups, one African (Eastern Africa) and other European (including American populations). In the present case, all the northern Mediterranean populations clustered together. When the dendrogram was produced for the Iberian populations alone, no clear clustering pattern was observed. In the Iberian Peninsula, olive groves cover huge expanses of territory. This, along with the data obtained in this study, strongly suggests the existence of a large northern Mediterranean olive fly population rather than several small and isolated populations. Gene flow and passive transport may be the reasons for the genetic similarity seen between the north Mediterranean populations. This information is of significant value in terms of the control of this pest.

In addition, the American population tested, along with other sample from a previous study, showed a greater degree of within-population similarity than similarity with other populations, indicating it to be the product of a single introduction event and later dispersion.

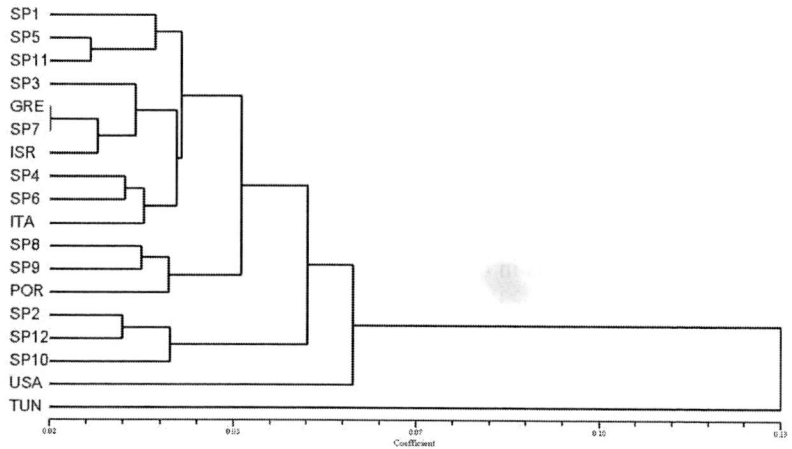

Figure 8. UPGMA dendrogram of *B. oleae* populations using Nei's genetic distances inferred from RAPD.

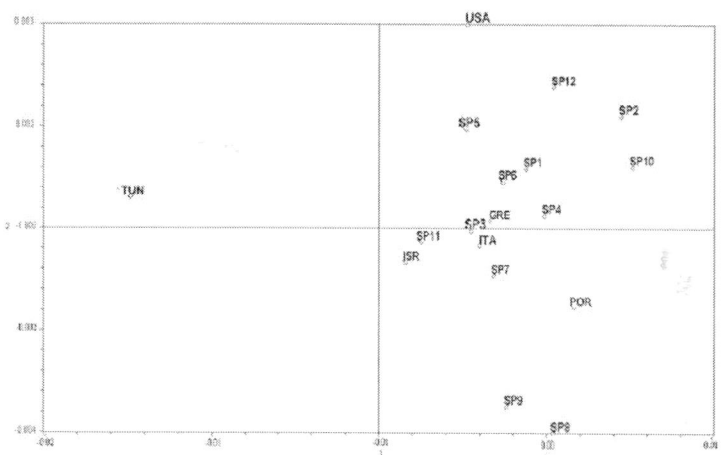

Figure 9. Plot of the first two components in a principal component analysis of RAPD data from the eighteen populations of olive fruit fly sampled.

In summary, phylogeographic studies involving RAPD markers have proven useful for characterizing the genetic structure of olive fly populations. A substantial level of polymorphism has been detected, along with considerable genetic diversity, though predominantly within populations (genetic diversity is low among populations). Strategies for future control programs should take this information into account. The low level of genetic differentiation among the north Mediterranean populations, and the important gene flow detected between them, show the need for integrated control programs coordinated between different geographical areas.

Chapter 8

CONCLUSION

The classical methods of plant protection, i.e., the indiscriminate use and abuse of pesticides, have proven to be increasingly unsustainable and cost-ineffective due to the development of pest resistance, the rising financial costs of pesticide use, and the negative effects of pesticide use on human health and the environment (FAO, http://www.fao.org/).

Although the need to develop and implement more effective strategies of combating pests and pathogens has always been dire, the urgency of this challenge has increased sharply in recent years due to the globalisation of trade, the increase in the human population and therefore the increased demand for food, and social awakening with respect to health and biodiversity. Consumers have made clear that they want safe, more effective pest management methods while being able to protect the environment and their own health. The consensus is that better scientific knowledge of insect pests and their natural enemies could lead to more effective biological control. Certainly, molecular methodologies can provide us with new and useful information in this and other respects. Important questions related to population structure and dynamics can be answered through phylogeographic analysis using different molecular markers and technologies. This is important since the control of pests requires very different strategies depending on the variation present in and between different populations. A single bioinvasion and later expansion may need to be dealt with differently than several or continuing bioinvasions, as might a pest that has recently colonised an area compared to one that is long established. The same is true for large, widespread populations compared to a number of different populations, large populations compared to those that suffer periodic bottlenecks, and those with high gene flow compared to isolated populations, etc.

Scientific knowledge is the cornerstone on which to build success in Integrated Pest Management, and agricultural associations are now demanding the EU legislate on a scientific basis. Phylogeographic knowledge *must* be taken into account if we are to a fight insect pests, in a way that is more efficient, more cost-effective, and safer for our health and environment.

REFERENCES

Abrahamson, WG; Weis, AE (1997). *Evolutionary ecology across three trophic levels: Goldenrods, Gallmakers, and Natural Enemies.* Monographs in Population Biology 29, Princeton University Press.

Afonso, JM; Prestano J; Hernández, M. Rapid isolation of mitochondrial DNA from *Drosophila* adults. *Biochemical Genetics*. 26, 381-386. 1988.

Alberola, TM; Aptosoglou, S; Arsenakis, M; Bel, Y; Delrio, G; Ellar, DJ; Ferre, J; Granero, F; Guttmann, DM; Koliais, S; Martinez-Sebastian, MJ; Prota, R; Rubino, S; Satta, A; Scarpellini, G; Sivropoulou, A; Vasara, E. Insecticidal activity of strains of *Bacillus thuringiensis* on larvae and adults of *Bactrocera oleae* Gmelin (Diptera, Tephritidae). *Journal of Invertebrate Pathology* 74, 127-136. 1999.

Aluja, M; Mangan, RL. Fruit Fly (Diptera: Tephritidae) host status determination: critical conceptual, methodological, and regulatory considerations. *Annual Reviews of Entomology* 53, 473-502. 2008.

Augustinus, AA; Mamuris, Z; Stratikopoulos, EE; D´Amelio, S; Zacharopoulou, A; Mathiopoulos, KD. Microsatellite analysis of olive fly populations in the Mediterranean indicates a westward expansion of the species. *Genetica* 125, 231-241. 2005.

Avise, JC. Mitochondrial DNA and the evolutionary genetics of higher animals. *Philosophical Transactions of the Royal Society London B* 312:325-342. 1986.

Avise, JC. Matriarchal liberation. *Nature* 352:192. 1991.

Avise, JC. Ten unorthodox perspectives on evolution prompted by comparative population genetic findings on mitochondrial DNA. *Annual Review of Genetics* 25:45-69. 1991.

Avise, JC. (1994). *Molecular Markers, Natural History, and Evolution.* Chapman & Hall, New York (511 pp.).

Avise, JC. (2000). *Phylogeography: The History and Formation of Species.* Harvard University Press, Cambridge, MA. (447 pp.).

Avise, JC. An American naturalist's impressions on Australian biodiversity and conservation. *Biodiversity and Conservation* 12:1-7. 2003.

Avise, JC. The best and the worst of times for evolutionary biology. *BioScience* 53:247-255. 2003.

Avise, JC. (2004). *Molecular Markers, Natural History, and Evolution* (Second Edition). Sinauer, Sunderland, MA. (684 pp.).

Avise, JC; Arnold, J; Ball Jr., RM; Bermingham, E; Lamb, T; Neigel, J.E; Reeb, CA; Saunders, NC. Intraspecific phylogeography: the mitochondrial DNA bridge between population genetics and systematics. *Annual Review of Ecology and Systematics* 18:489-522. 1987.

Barranco, D; Fernández-Escobar, R; Rallo, L; Eds. (2004). *El cultivo del olivo.* Mundi-Prensa-Junta de Andalucia, España.

Baruffi, L; Damian, G; Guglielmino, CR; Bandi, C; Malacrida, AR; Gasperi, G. Polymorphism within and between populations of *Ceratitis capitata*: comparison between RAPD and multilocus enzyme electrophoresis data. *Heredity* 74, 425-437. 1995.

Behura, SK. Molecular marker systems in insects: current trends and future avenues. *Molecular ecology.* 15(11): 3087. 2006.

Beitia, F. Nueva especie de mosca blanca en las Islas Canarias. *Terralia*, 4: 24-25. 1998.

Bellows, TS; Fisher, TW eds.. (1999). *Handbook of Biological Control. Principles and applications of biological control*, Academic Press, New York, USA.

Bohonak, AJ; Davies, N; Villablanca, FX; Roderick, GK. Invasion genetics of New World medflies: testing alternative colonization scenarios. *Biological Invasions* 3: 103-111. 2001.

Bonizzoni, M; Malacrida, A.R; Guglielmino, CR; Gomulski, LM; Gasperi, G; Zheng, L. Microsatellite polymorphism in the Mediterranean fruit fly, *Ceratitis capitata. Insect Molecular Biology* 9, 251-261. 2000.

Bonizzoni, M; Zheng, L; Guglielmino, CR; Haymer, DS; Gasperi, G; Gomulski, LM; Malacrida, AR. Microsatellite analysis of medfly bioinfestations in California. *Molecular Ecology* 10:2515–2524. 2001.

Bonizzoni, M; Guglielmino, CR; Smallridge, CJ; Gomulski, LM; Malacrida, AR; Gasperi, G. On the origins of medfly invasion and expansion in Australia. *Molecular Ecology* 13: 3845-3855. 2004.

Bush, GL; Kitto, GB. Research on the genetic structure of wild and laboratory strains of the olive fly. FAO Report: *Development of pest management system for olive culture program.* Food and Agricultura Organization of the United Nations, Rome. 1979.

Callejas, C; Roda, P; Reyes, A; Ochando, MD. Identificación genética de *Dacus –Bactrocera- oleae* Gmelin (Diptera: Tephritidae) mediante marcadores RAPD-PCR. *Boletin de Sanidad Vegetal y Plagas* 24, 873-882. 1998.

Callejas, C; Ochando, MD. Allozymic variability in Spanish populations of *Ceratitis capitata. Fruits* 59, 181-190. 2004.

Callejas, C; Gobbi, A; Velasco, A; Beitia, FJ; Ochando, MD. The use of RAPD markers to detect genetic patterns in *Aleurodicus dispersus* (Hemiptera: Aleyrodidae) populations from the Canary Islands. *European Journal of Entomology*, 102, 289-291. 2005.

Carey, JR. Establishment of the Mediterranean fruit fly in California. *Science* 253, 1369-1373. 1991.

Coombs, J; Coombs, R. (2003). *A dictionary of Biological Control and Integrated Pest Management* (3rd Edition), 300 pp. CPL Press, Newbury, UK.

Cox, JSTH. (2007). The role of geographic information systems and spatial analysis in Area-Wide vector control programmes, pp.199-210, In *"Area-Wide control of Insect pests"*, Vreysen et al., eds., Springer, Dordrecht, The Netherlands.

Davies, N; Villablanca, FX; Roderick, GK. Bioinvasions of the Medfly *Ceratitis capitata*: source estimation using DNA sequences at multiple intron loci. *Genetics* 153: 351-360. 1999.

De Sousa, GB; de Dutari, GP; Gardenal, CN. Genetic structure of *Aedes albifasciatus* (Diptera: Culicidae) populations in Central Argentina determined by random amplified polymorphic DNA-polymerase chain reaction markers. *Journal of Medical Entomology* 36, 400-404. 1999.

Debach, P. (1964). *Biological control of insect pest and weeds.* Reinhold, New York, USA. 844 pp.

Dowell, RV; Krass, CJ. Exotic pests pose growing problem for California. *California Agriculture* 46: 6–12. 1992.

Downie, DA; Fisher, JR; Granett, J. Grapes, galls, and geography: the distribution of nuclear and mtDNA variation across host plant species and regions in a specialist herbivore. *Evolution* 55, 1345-1362. 2001.

European Environment Agency (EEA), Office for Official Publications of the European Communities (OPOCE). *Europe's environment — The fourth assessment*. European Environment Agency. Report No 1/2007. Copenhagen, Denmark.

Felsenstein, J. (2004). *Inferring Phylogenies*. Sinauer Associates, Sunderland, Massachusetts, USA.

Fimiani, P. (1989) Mediterranean region. pp. 37-55 *in* Robinson, A.S. & Hooper, G.H. (Eds.) *Fruit flies: Their Biology, Natural enemies and Control*. Vol. 3A. Elsevier, Amsterdam, The Netherlands.

Fletcher, B.S. (1989). Life history strategies of tephritid fruit flies. In *Fruit flies: their Biology, Natural Enemies and Control*, Vol 3A, eds. Robinson and Hooper: 195-208. Elsevier, Amsterdam.

Gallo, D.N.O., Wiendel, F.M., Silvera Neto, S. & Ricardo, P.L.C. (1970) *Manual de Entomologia Agronomica*. Ceres, São Paulo, Brasil.

Gasperi, G; Guglielmino, CR; Malacrida, AR; Milani, R. Genetic variability and gene flow in geographical populations of *Ceratitis capitata* (Wied.) (medfly). *Heredity* 67: 347-356. 1991.

Gasperi, G; Bonizzoni, M; Gomulski, LM; Murelli, V; Torti, C; Malacrida, AR; Guglielmino, CR . Genetic differentiation, gene flow and the origin of infestations of the medfly, *Ceratitis capitata*. *Genetica* 116: 125–135. 2002.

Gasparich, GE; Sheppard, WS; Han, HY; McPheron, BA; Steck, GJ. Analysis of mitochondrial DNA and development of PCR-based diagnostic molecular markers for the Mediterranean fruit fly (*Ceratitis capitata*) populations. *Insect Molecular Biology* 4, 61-67. 1995.

Gasparich, GE; Silva, JG; Han, HY; McPheron, BA; Steck, GJ; Sheppard, WS. Population genetic structure of Mediterranean Fruit Fly (Diptera: Tephritidae) and Implications for Worldwide Colonization Patterns. *Annals of the Entomological Society of America* 90: 790-797. 1997.

Gomulski, LM; Bourtzis, K; Brogna, S; Morandi, PA; Bonvicini, C; Sebastiani, F; Torit, C; Guglielmino, CR; Savakis, C; Gasperi, G; Malacrida, AR. Intron size polymorphism of the *Adh1* gene parallels the worldwide colonization history of the Mediterranean fruit fly, *Ceratitis capitata*. *Molecular Ecology* 7: 1729-1741. 1998.

Greathead, DJ; Greathead AH . Biological control of insect pests by insect parasitoids and predators: the BIOCAT database. *Biocontrol News Information* 13: 61N–68N. 1992.

Hagen, KS; William, WW; Tassan, RL. Mediterranean fruit fly: The worst may be yet to come. *California Agriculture* (University of California, Division of Agricultural Sciences, Reports of progress in research), March-April 1981, 35: 5–7. 1981.

Haymer, S; McInnis, DO. Resolution of populations of the Mediterranean fruit fly at the DNA level using random primers for the polymerase chain reaction. *Genome* **37**: 244-248. 1994.

He, M; Haymer, S. Genetic relationships of populations and the origins of new infestations of the Mediterranean fruit fly. *Molecular Ecology* 8, 1247-1257. 1999.

Headrick, DH; Goeden, RD. Issues concerning the eradication or establishment and biological control of the Mediterranean fruit fly, *Ceratitis capitata* (Wiedemann) (Diptera: Tephritidae) in California. *Biological Control* **6**, 412-421. 1996.

Heckel, DG. Genomics in pure and applied entomology. *Annual Review of Entomology* 48, 235-260. 2003.

Hernández-Suárez, E; Carnero, A; Hernández, M; Beitia, F; Alonso, C. *Lecanoideus floccissimus* (Homoptera, Aleyrodidae): Nueva plaga en las Islas Canarias. *Phytoma-ESPAÑA* 91: 35-48. 1997.

Hernández-Suárez E. (1999). *La familia Aleyrodidae y sus enemigos naturales en las Islas Canarias*. PhD Thesis, Universidad de La Laguna (Tenerife, Islas Canarias), 687pp.

Higuchi, R., 1989, Simple and rapid preparation of samples for PCR, in H.A. Erlich (ed) *PCR Technology* (New York: Stockton Press), pp: 31-38.

Hillis, DM; Moritz, C; Mable, BK. (1996) *Molecular Systematics*. 2 edition. Sinauer Associates Inc., Sunderland, MA, USA.

Huettel, MD; Fuerst, PA; Maruyama, M; Chakraborty, R. Genetic effects of multiple population bottlenecks in the Mediterranean fruit fly (*Ceratitis capitata*). *Genetics* 94s, 47 (abstract). 1980.

Kourti, A. Comparison of mtDNA variants among Mediterranean and New World introductions of the Mediterranean fruit fly *Ceratitis capitata* (Wied.). *Biochemical Genetics* **35**, 363-370. 1997.

Kourti, A. Patterns of variation within and between Greek populations of *Ceratitis capitata* suggest extensive gene flow and latitudinal clines. *Journal of Economic Entomology* 93, 1186-1190. 2004.

Lin, H; Downie, DA; Walker, MA; Granett, J; English-Loeb, G. Genetic structure in native populations of grape phylloxera (Homoptera: Phylloxeridae). *Annals of the Entomological Society of America* 92, 376-381. 1999.

Lomolino, MV; Riddle, BR; Brown, JH. (2006). *Biogeography*. Sinauer Assocciates, Inc. Sunderland, Massachusetts, USA.

Loukas, M. (1989) Population genetic studies of fruit flies of economic importance, specially medfly and olive fruit fly, using electrophoretic methods. pp. 69-102 *in* Loxdale, H.J. & den Hollander, J. (Eds.)

Electrophoretic Studies on Agricultural Pests. Clarendon Press, Oxford, United Kingdom.

Loxdale, HD; Lushai, G. Molecular markers in entomology (Review). *Bulletin of Entomological Research* 88, 577 – 600. 1998.

Malacrida, AR; Guglielmino, CR; Gasperi, G; Baruffi, L; Milani, R. Spatial and temporal differentiation in colonizing populations of *Ceratitis capitata. Heredity* 69, 101-111. 1992.

Malacrida, AR; Marinoni, F; Tori, C; Gomulski, LM; Sebastián, F; Bonvicini, C; Gasperi, G; Guglielmino, CR. Genetic aspects of the worldwide colonization process of *Ceratitis capitata. The Journal of Heredity* 89, 501-507. 1998.

Malacrida, AR; Gomulski, LM; Bonizzoni, M; Bertin, S; Gasperi, G; Guglielmino, CR . Globalization and fruit fly invasion and expansion: the medfly paradigm. *Genetica* 131: 1–9. 2007.

Martin, JH; Mifsud, D; Rapisarda, C. The whiteflies (Hemiptera:Aleyrodidae) of Europe and the Mediterranean Basin. *Bulletin of Entomological Reserach* 90: 407-448. 2000.

Mayurama, T. Effective number of alleles in a subdivided population. *Theoretical Population Biology* 1, 273-306. 1970.

Mayurama T. Distribution of gene frequencies in a geographically structured finite population. I. Distribution of neutral genes and of genes with a small effect. *Annals of Human Genetics* 35, 411-423. 1972.

Mcpheron, BA; Gasparich, GE; Han, H-Y; Steck, GJ; Sheppard, WS. Mitochondrial DNA restriction map for the Mediterranean fruit fly, *Ceratitis capitata. Biochemical Genetics* 32: 25-33. 1994.

Meixner, MD; McPheron , BA; Silva , JG; Gasparich , GE; Sheppard, WS . The Mediterranean fruit fly in California: evidence for multiple introductions and persistent populations based on microsatellite and mitochondrial DNA variability. *Molecular Ecology* 11: 891–899. 2002.

Mendelson, TC; Shaw, KL. Use of AFLP markers in surveys of arthropod diversity. *Methods in Enzymology*, 395: 161-177. 2005.

Nardi, F; Carapelli, A; Dallai, R; Roderick, GK; Frati, F. Population structure and colonization history of the olive fly, *Bactrocera oleae* (Diptera, Tephritidae). *Molecular Ecology* 14, 2729-2738. 2005.

Nei, M; Li, WH. Mathematical model for studying genetic variation in terms of restriction endonucleases. *Proceedings of the National Academy of Sciences. USA* 76, 5269-5273. 1979.

Nei, M; Tajima, F. DNA polymorphism detectable by restriction endonucleases. *Genetics* 97, 145-163. 1981.

Nei, M; Kumar, S. (2000) *Molecular Evolution and Phylogenetics*. Oxford University Press, New York, USA.

Ochando, MD; Callejas, C; Fernández, OH; Reyes, A. Variabilidad genética aloenzimática en *Dacus oleae* (Gmelin) (Díptera: Tephritidae) I. Análisis de dos poblaciones naturales del sureste español. *Boletín de Sanidad Vegetal y Plagas*, 20, 35-44. 1994.

Ochando, MD; Reyes, A. Genetic population structure in olive fly *Bactrocera oleae* (Gmelin): gene flow and patterns of geographic differentiation. *Journal of Applied Entomology* 124, 177-183. 2000.

Ochando, MD; Reyes, A.; Callejas, C; Segura, D and Fernández, P. Molecular genetic methodologies applied to the study of fly pests. *Trends in Entomology*, 3, 73-85. 2003a.

Ochando, MD; Reyes, A; Callejas, C. Genetic structure of *Ceratitis capitata* species: within and between population variability. *Integrated Control in Citrus Fruit Crops. IOBC/wprs Bulletin* 26 (6): 59-72. 2003b.

Ochando, MD; Beroiz, B; Callejas, C; Hernández-Crespo, P; Ortego, F; Castañera, C. Genética poblacional de *Ceratitis capitata* mediante el empleo de marcadores moleculares. *Levante Agrícola*, 385, 139-144. 2007.

Reyes, A. (1995) Análisis de la variabilidad genética en poblaciones españolas de *Ceratitis capitata* Wied. mediante la utilización de marcadores moleculares. Ph. D. Thesis. Complutense University, Madrid, Spain, pp. 202.

Reyes, A; Ochando, MD. A study of gene-enzyme variability in three Spanish populations of Ceratitis capitata. *International Organization for Biological Control. West Palaeartic Regional Section Bulletin* 17, 151-160. 1994.

Reyes, A; Linacero, R; Ochando, MD. Molecular genetics and integrated control: A universal genomic DNA microextraction mehtod for PCR, RAPD, restriction and Southern análisis. *IOBC/wprs Bulletin* 20 (4): 274-284. 1997.

Reyes, A; Ochando, MD. Genetic differentiation in Spanish populations of *Ceratitis capitata* as revealed by abundant soluble proteins. *Genetica* 104, 59-66. 1998a.

Reyes, A; Ochando, MD. Use of molecular markers for detecting the geographical origin of *Ceratitis capitata* (Diptera:Tephritidae) populations. *Annals of the Entomological Society of America* 91, 222-227. 1998b.

Reyes, A; Ochando, MD. Mitochondrial DNA variation in Spanish populations of *Ceratitis capitata* (Wiedeman) (Tephrtidae) and the colonization process. *Journal of Applied Entomology* 128, 358-364. 2004.

Rice, RE. Bionomics of the olive fruit fly *Bactrocera (Dacus) oleae*. Plant Protection Quaterly 10, 1-5. 2000.

Rice, RE; Phillips, PA; Stewart-Leslie, J; Sibbett, GS. Olive fruit fly populations measured in central and southern California. *California Agriculture* 57, 122-127. 2003.

Robinson, AS; Hooper, G. (1989). *Fruit flies, their biology, natural enemies and control.* Elsevier Amsterdam, The Netherlands.

Roderick, GK; Navajas, M. Genotypes in novel environments: Genetics and evolution in biological control. *Nature Reviews Genetics* 4, 889-899. 2003.

Ruiz, A. (1948). *Fauna entomológica del olivo en España. Estudio sistemático y biológico de las especies de mayor importancia económica.* Trabajos del Instituto Español de Entomología. Madrid, Spain.

Sambrook, J; Fritsch, EF; Maniatis, T. (1989). *Molecular cloning: a laboratory manual.* 2nd edit. Cold Spring Harbor Laboratory. Cold Spring Harbor, New York, USA.

Segura, D; Callejas, C; Ochando, MD. Molecular markers as useful tools for population genetics of the olive fly, *Bactrocera oleae. IOBC wprs Bulletin,* 30, 79-87. 2007.

Segura, D; Callejas, C; Ochando, MD. *Bactrocera oleae*: a single large population in the Northern Mediterranean basin. *Journal of Applied Entomology,* 132, 706-713. 2008.

Severson, SE; Brown, B; Knudson, DL. Genetic and physical mapping in mosquitoes: molecular approaches. *Annual Review of Entomology* 46, pp. 183–219. 2001.

Sheppard, WS; Steck, GJ; McPheron, BA. Geographic populations of the medfly may be differentiated by mitochondrial DNA variation. *Experientia* 48, 1010-1013. 1992.

Silva, J; Mexner, M; McPheron, B; Steck, G; Sheppard, W. Recent Mediterranean fruit fly (Diptera: Tephritidae) infestations in Florida. A genetic perspective. *Journal of Economic Entomology* 96, 1711–1718. 2003.

Steck, GJ; Gasparich, GE; Han, HY; McPheron, BA; Sheppard, WS. Distribution of mitochondrial DNA haplotypes among *Ceratitis capitata* populations worldwide. pp. 291-296 *in* McPheron, B.A. & Steck, G.J. (Eds.) *Fruit fly pests. A world assessment of their biology and management.* St. Lucie Press, Delray Beach, FL, USA. 1996

Swofford, DL; Olsen, GJ; Waddell, PJ; Hillis, DM. (1996). Phylogenetic inference pp. 407-514, in *Molecular Systematics*, Hillis, D.M., Moritz, C., Mable, B.K. eds., Sinauer Assoc., Sunderland, MA, USA.

Tan, K-H ed. (2000). *Area-wide control of fruit flies and other insect pests.* IAEA. *Sinaran Bros. Sdn. Bhd* Penerbit University Sains Malasia, 782 pp.

Unruh, TR; Woolley, JB. (1999). Molecular methods in classical biological methods. In *"Handbook of Biological Control"*, Bellows and Fisher, eds., Academic Press, New York, USA.

Villablanca, FX; Roderick , GK; Palumbi, SR . Invasion genetics of the Mediterranean fruit fly: variation in multiple nuclear introns. *Molecular Ecology* 7: 547–560. 1998.

Vreysen, MJ B; Robinson, AS; Hendrichs, J. (Eds.) 2007. *Area-wide Control of Insect Pests: From Research to Field Implementation.* Springer, Dordrecht, The Netherlands. 789 pp.

Waterhouse, DF; Norris, KR. (1989). *Aleurodicus dispersus* spiraling whitefly. pp. 12-23. In: Biological Control Pacific Prospects - Supplement 1. Australian Center for International Agriculture Research, Canberra, Australia.

Welsh, J; McClelland, M. Fingerprinting genomes using PCR with arbitrary primers. *Nucleic Acids Research* 18, 7213-7218. 1990.

Williams, JKG; Kubelik, AR; Livak, KJ; Rafalsky, JA; Tyngey, SV. DNA polymorphisms amplified by arbitrary primers are useful as genetic markers. *Nucleic Acids Research* 18, 6531-6535. 1990.

Wright, S. Evolution in Mendelian populations. *Genetics* 16, 97-159. 1931.

Yong. HS. Allozyme variation in the melon fly *Dacus cucurbitae* (Insecta: Diptera: Tephritidae) from Peninsular Malaysia. *Comparative Biochemistry and Physiology* 102B, 367-370. 1992.

Zhang, DX; Hewitt, GM. Insect mitochondrial control region: a review of its structure, evolution and usefulness in evolutionary studies. *Biochemical Systematics and Ecology* 25: 99-120. 1997.

Zitoudi, K; Margaritopoulos, JT; Mamuris, Z; Tsitsipis, JA. Genetic variation in *Myzus persicae* populations associated with host-plant and life cycle category. *Entomologia Experimentalis et Applicata* 99, 303-311. 2001.

Zouros, E; Loukas, M. (1989). Biochemical and colonization genetics of *Dacus oleae* (Gmelin). In: *Fruit flies, Their biology, Natural enemies and Control.* A. S. Robinson and G. Hooper (eds.). Chapter 5.3, 75-87.

INDEX